THE PLANETS OF OUR SOLAR SYSTEM

VENUS

by Jane Parks Gardner

BrightPoint Press

San Diego, CA

BrightPoint Press

an imprint of ReferencePoint Press, Inc.
Printed in the United States

For more information, contact:
BrightPoint Press
PO Box 27779
San Diego, CA 92198
www.BrightPointPress.com

LIBRARY OF CONGRESS CATALOGING-IN-PUBLICATION DATA

Name: Parks Gardner, Jane, author.
Title: Venus / by Jane Parks Gardner.
Description: San Diego, CA: ReferencePoint Press, 2026 | Series: The planets of our solar system | Audience: Grade 7 to 9 | Includes bibliographical references and index.
Identifiers: ISBN: 9781678211769 (hardcover) | ISBN: 9781678211776 (eBook)
The complete Library of Congress record is available at www.loc.gov.

CONTENTS

AT A GLANCE

- Venus is the second planet from the Sun. It orbits between Mercury and Earth.
- Venus is one of four terrestrial planets in the solar system. This means it has a rocky surface.
- Although it is not the closest planet to the Sun, Venus is the hottest planet in the solar system. This is because its dense atmosphere traps heat.
- Venus is one of the brightest objects in the night sky. People have observed it since before recorded history.
- The invention of the telescope in the 1600s allowed people to study Venus in greater detail. Ground-based radar technology help people learn even more in the 1900s.

- People have sent many spacecraft to explore Venus. *Mariner 2* performed the first successful flyby of Venus in 1962.

- The Soviet Union's Venera program sent several spacecraft to Venus between 1961 and 1983. Some of these spacecraft landed on Venus's surface to study the planet up close.

- Venus has no moons. Astronomers have discovered a small space rock that appears to be orbiting Venus, but it actually orbits the Sun.

- Unlike Earth, Venus does not have a strong magnetic field.

LANDING ON VENUS

On October 30, 1981, the Soviet Union launched a spacecraft toward Venus. It was called *Venera 13*. A rocket carried the spacecraft out of Earth's orbit. *Venera 13* traveled through space for 4 months. Then it finally reached Venus. The spacecraft fell through the planet's thick clouds. It released a parachute that slowed it down.

Venera 13 successfully landed on Venus's surface. Conditions on the planet

This illustration shows *Venera 13* conducting scientific research on Venus's surface.

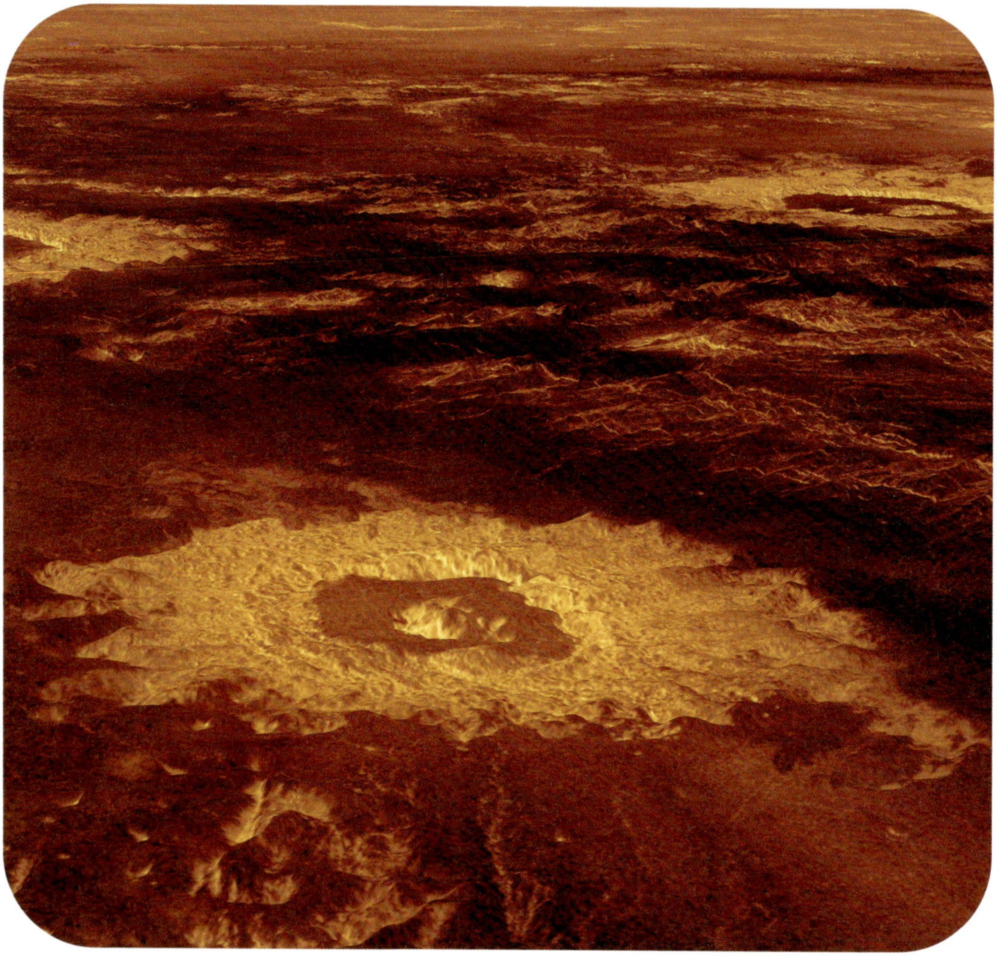

Craters scar Venus's harsh surface.

were incredibly intense. The air was so thick that the pressure was about ninety times what it is on Earth. The heat was hot enough to melt lead. Clouds of sulfuric acid swirled overhead. Volcanoes rose from vast plains of hardened lava.

This harsh environment is hostile to life as people understand it. Not even *Venera 13* could withstand it for long. After 2 hours, scientists on Earth stopped receiving signals from the lander.

Venera 13 studied Venus's surface in the short time it had. It drilled into the planet's soil and took samples. It took photos of its surroundings. *Venera 13* was not the first spacecraft to land on Venus. But it was the first to send color photos of the surface back to Earth. These pictures have amazed and inspired people.

AN EXTREME WORLD

Venus might seem a lot like Earth at first glance. They are almost the same size. They are both made of rocks and metals.

And both have a substantial atmosphere.

But Venus is unlike Earth in other ways.

A volcano rises from Venus's surface in this computer-generated image with exaggerated vertical scale.

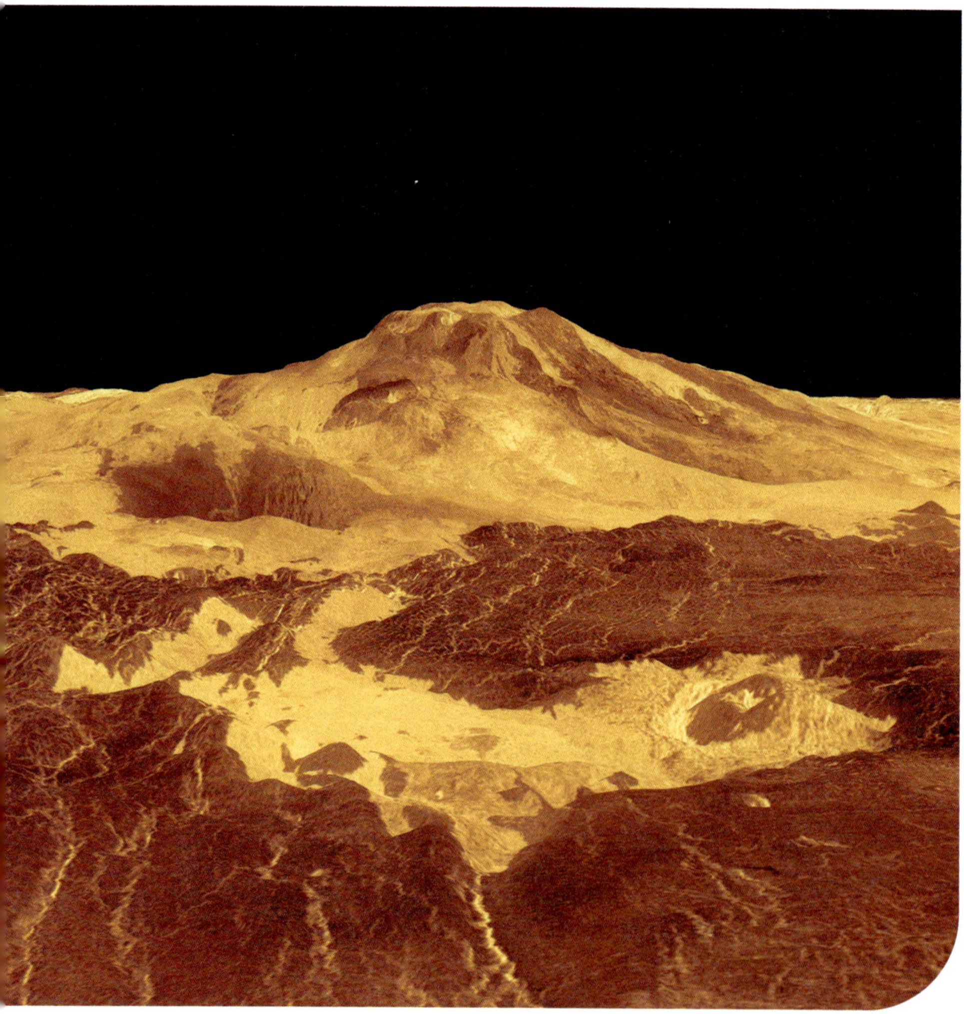

Venus is a dangerous world beneath its thick, yellow clouds.

Scientists have sent spacecraft to peer through Venus's dense clouds and map its surface. What these spacecraft have found is astonishing. Venus has massive volcanoes and strange craters. Scientists have learned a lot from studying Venus's harsh environment.

MEET VENUS

Venus is the second planet from the Sun. It orbits the Sun between Mercury and Earth. Venus is about 67 million miles (108 million km) away from the Sun on average. Venus's orbit is nearly circular. This means that it is about the same distance from the Sun at all times.

Like Earth, Venus is a terrestrial planet. These planets have rocky surfaces. And they typically have similar internal structures.

The Sun shines on Venus, revealing its active atmosphere.

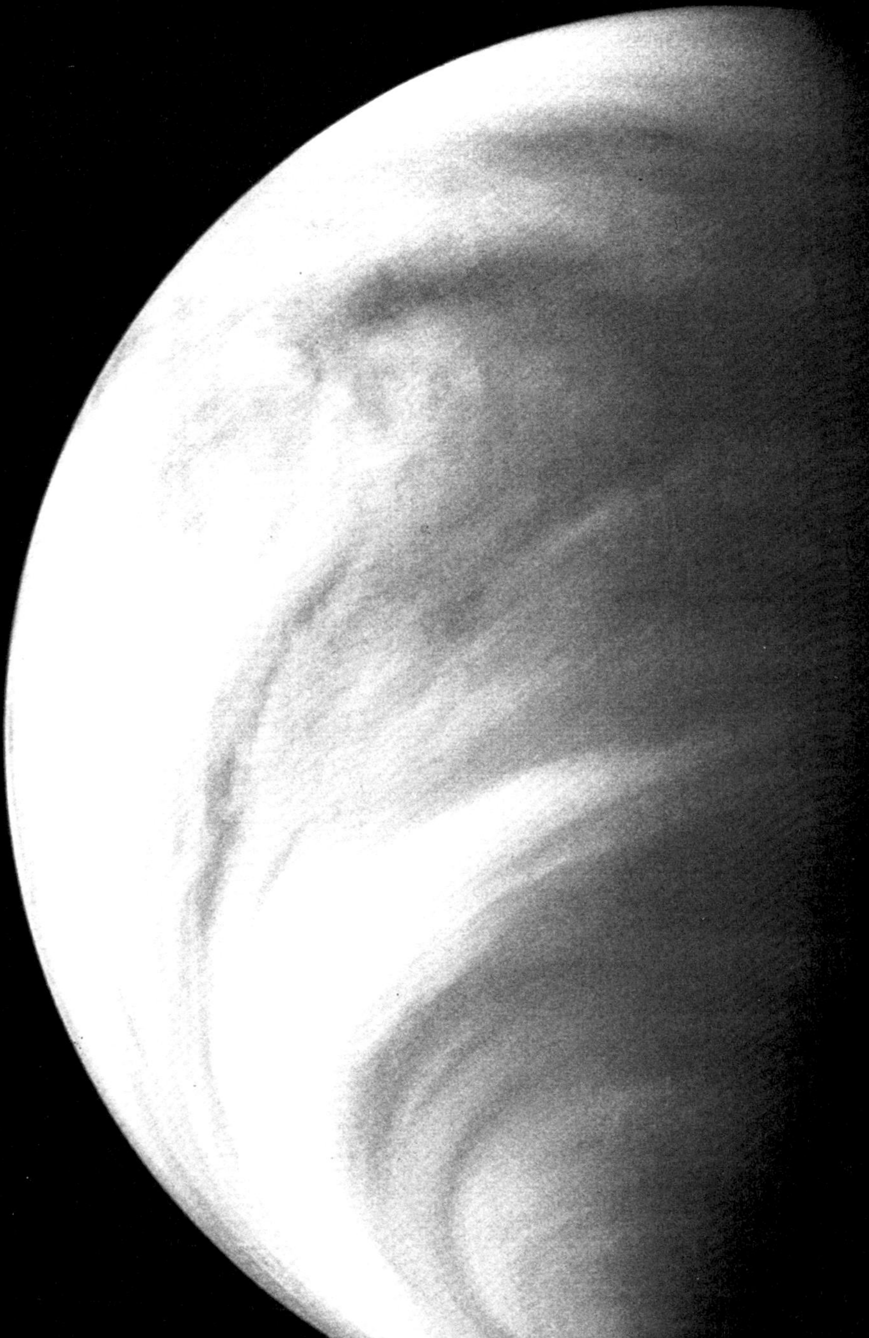

At the center of Venus is a metallic core. Surrounding the core is a molten layer called a mantle. The outer layer is called the crust. Venus's diameter at its equator is

The surface of a planet is the outer face of its crust.

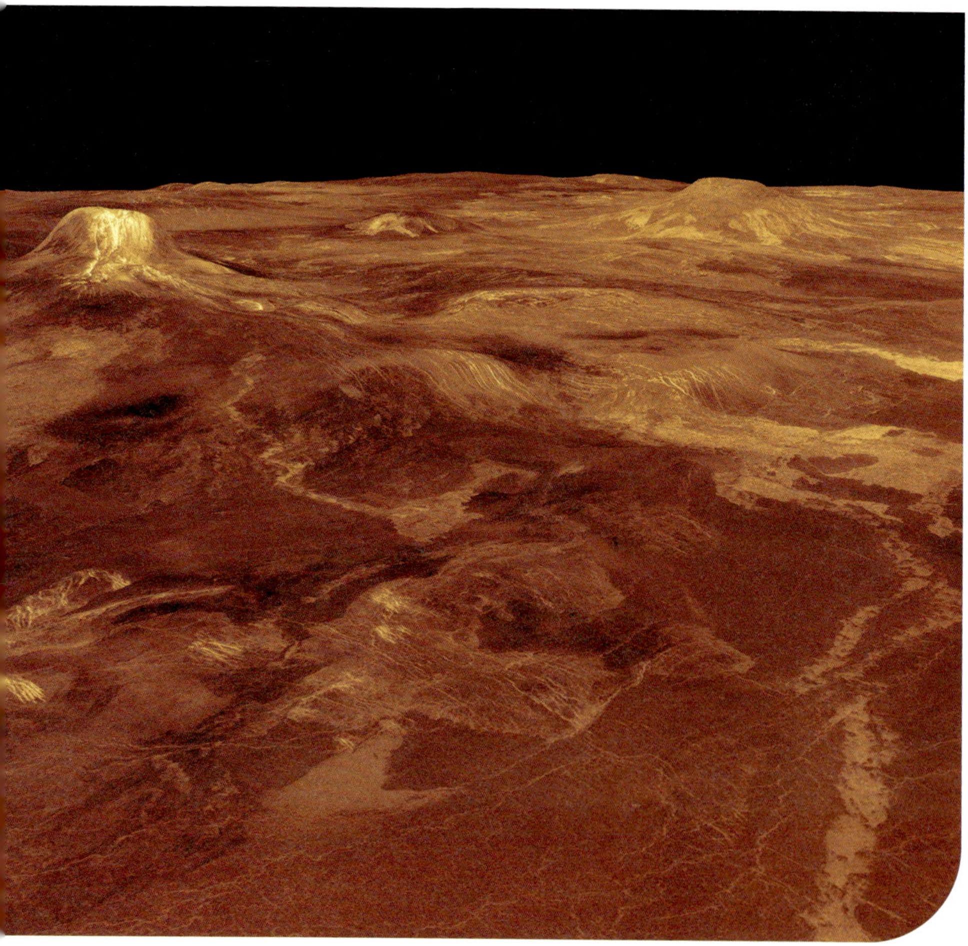

about 7,521 miles (12,104 km). This makes it slightly smaller than Earth.

Venus's **axis of rotation** is not very tilted. The axis is tilted away from the plane of Venus's orbit by only about 2.6 degrees. The top and bottom halves of the planet get roughly equal amounts of sunlight throughout the year. This means Venus does not have seasons.

A day on Venus is longer than its year. It takes about 243 Earth days for Venus to spin around its axis once. And it takes the planet only about 225 Earth days to orbit the Sun. Venus also spins in the opposite direction of most planets. This kind of motion is called retrograde. This means the Sun rises in the west and sets in the east on Venus.

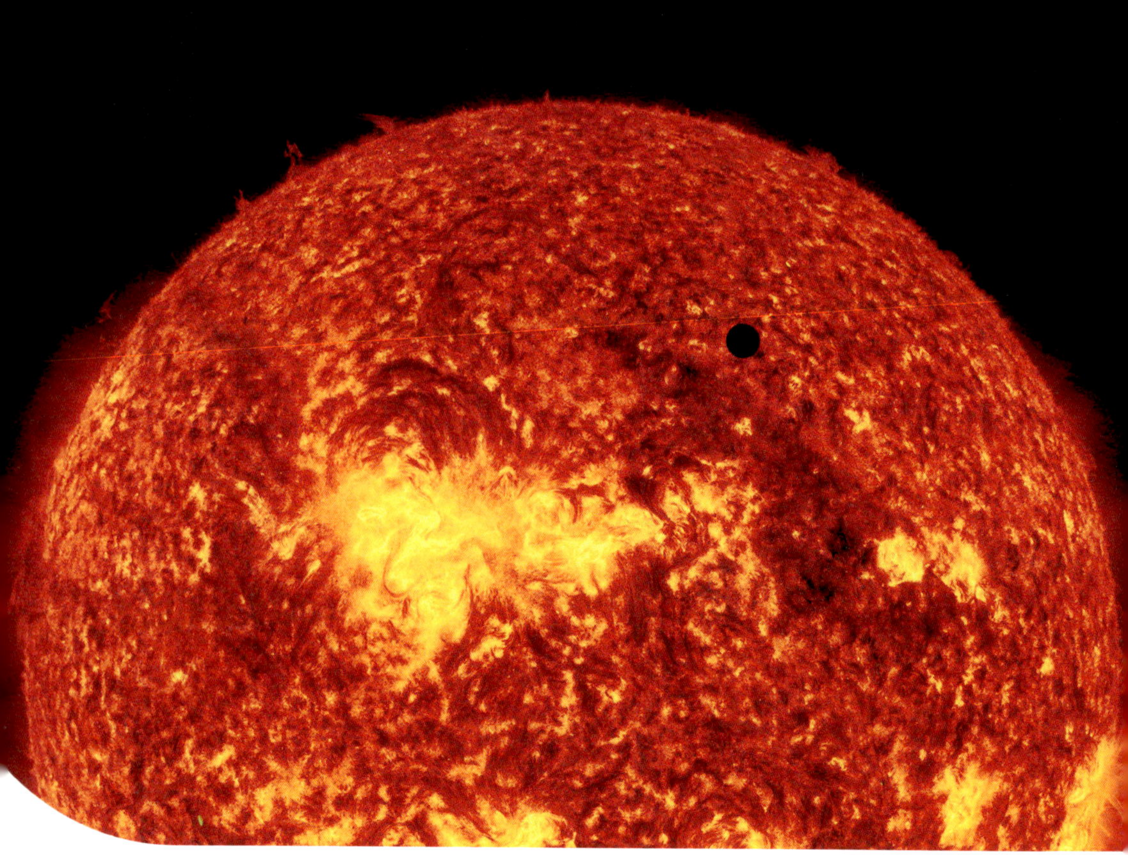

Venus passes between the Sun and Earth, appearing as a black dot against the bright star.

Venus's slow, retrograde rotation is unique in the solar system. Scientists wonder about what might have happened in the planet's past to cause its present conditions. One theory holds that Venus was struck by a large space rock early in its history. This may have reversed the direction of the planet's spin.

VENUS'S ATMOSPHERE

Venus is the hottest planet in the solar system. Its surface is about 867°F (464°C) on average. This makes Venus even hotter than Mercury, which is closer to the Sun. Venus's extreme temperature comes from its dense atmosphere. This layer of gases surrounding the planet traps heat in a process called the **greenhouse effect**. Earth also has a greenhouse effect. But Venus's is much stronger.

Venus's dense air is made mostly of carbon dioxide. The atmosphere also contains small amounts of sulfuric acid. These substances give Venus its yellowish color. The thick atmosphere creates huge pressure on the surface. The pressure on Venus is about ninety times greater than on Earth. **Astronomer** Stephen Kane explains, “Standing on the surface of Venus would be like standing at the bottom of a very hot ocean. You couldn’t breathe on it.”[1]

There is almost no water in Venus’s air. Scientists think **solar wind** carried the planet’s water away. Solar wind hit Venus’s upper atmosphere over billions of years. This pushed any water into space.

Even though Venus does not have water, it does have rain. Its rain is made of liquid sulfuric acid. The rain forms in Venus’s atmosphere and begins to fall. The rain becomes a gas before it can strike the planet’s surface. Venus also has strong winds. The fastest winds exist in the upper atmosphere. Wind speeds there can reach up to 224 miles per hour (360 kmh).

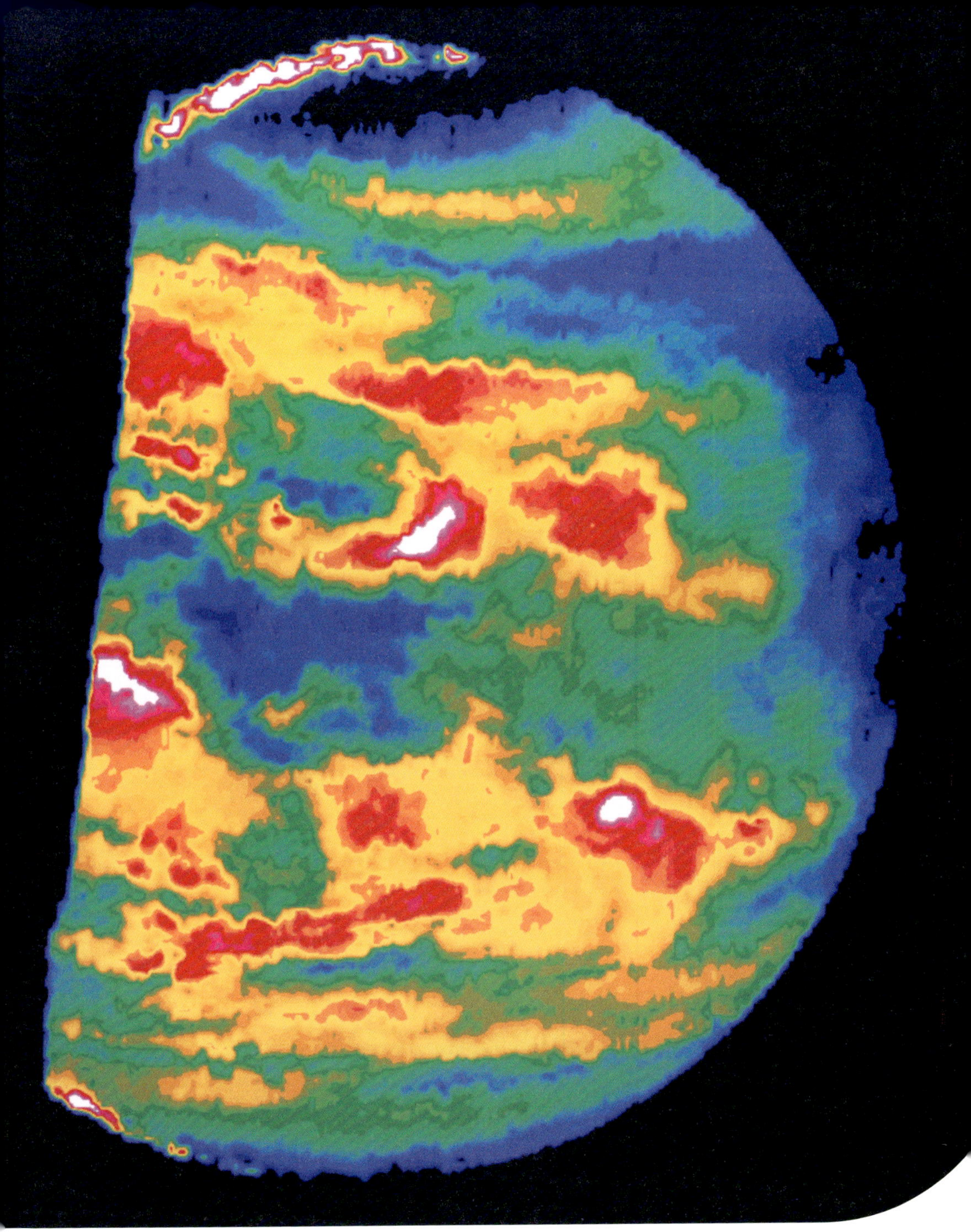

A spacecraft passing Venus captured this false-color image of the planet's clouds. The white and red areas represent places where clouds are thin.

BENEATH THE CLOUDS

Venus has a strange surface under its thick clouds. Scientist Giada N. Arney explains,

Basalt is a kind of igneous rock. Igneous rocks form by the cooling of molten rocks.

"Most of the surface of Venus is made of basalt, which is produced by volcanism."[2] Basalt is a dark rock that forms on a planet's surface from cooling lava. Basalt also forms on Earth. On Venus, thousands of volcanoes have shaped the planet's surface. Vast rocky plains stretch between volcanoes and mountains. The planet has no rivers or oceans.

Maxwell Montes is one mountain range on Venus. Its biggest mountain stands about 7 miles (11 km) tall at its peak. That is higher than Earth's Mount Everest. Maxwell Montes is part of a larger highland region on Venus called Ishtar Terra. This region is about as large as Australia. It is one of three major highland regions on the planet. The other two are called Aphrodite Terra and Lada Terra. Stretching between these highlands are lower lava plains.

Venus also has round features called coronae. These form when molten rock bubbles up to the surface through volcanoes. Some volcanoes on Venus may still be active today.

Venus's volcanism may explain why the planet's surface is relatively young.

Craters on Venus

Space rocks that have struck Venus have left their marks on its surface. Bowl-shaped craters appear across the planet's landscape. But craters smaller than about 1 mile (1.6 km) have not been found. This is because the space rocks that would form such craters break up in Venus's dense atmosphere.

Scientists estimate that the material on the planet's surface settled only 500 million years ago. They make these estimates by observing how many craters are on the surface. If the surface does not change, it should be covered in lots of old craters. But the relatively small number of craters on Venus suggests that the surface is young. Natural processes have reshaped the landscape.

CHAPTER TWO

STUDYING VENUS FROM EARTH

Venus has fascinated people for thousands of years. It is one of the brightest objects in the sky. Many stories have been inspired by this glowing planet. Various cultures have considered it a symbol of love, beauty, or war.

Venus is named after the Roman goddess of love and beauty. The ancient Greeks called it Aphrodite after their own goddess of love. These cultures also

Venus is the bright spot in the night sky in this image.

gave Venus and the other known planets symbols. Venus's symbol is commonly used today to represent the concepts of woman and female. Other cultures also gave the planet special names. In Babylon, Venus was called Ishtar. This was the goddess of fertility and war. One traditional name for Venus in China is Great White Golden Star.

Venus is often most visible at sunrise or sunset. Some ancient people thought Venus was two different objects. Ancient Greeks called its morning appearance Phosphorus. This means "bringer of light." They called its evening appearance Hesperus. This means "star of the evening." They later learned the two objects were the same.

PLANETARY SYMBOLS

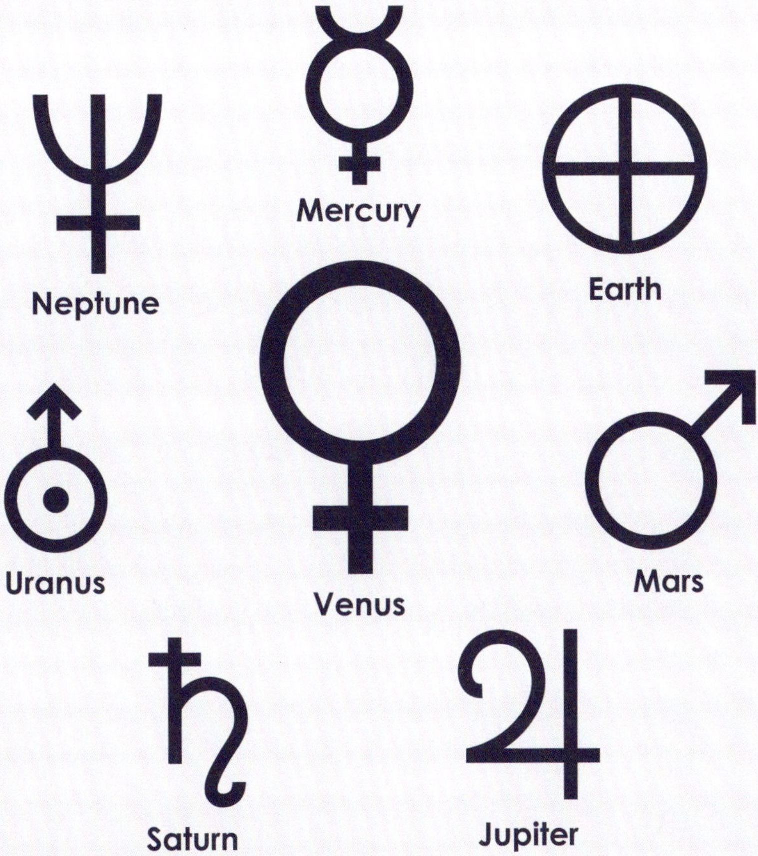

Each of the solar system's eight planets has a traditional symbol. However, modern astronomers rarely use these symbols.

OBSERVING VENUS FROM EARTH

The invention of the telescope helped people learn more about Venus. Galileo Galilei used one of the first telescopes to observe the planet in 1610. He saw that Venus has **phases** like the Moon. People in Galileo's time debated the structure of the universe. Some thought that Earth was the center of the universe. A newer idea claimed that the Sun was the center. Galileo's observation of Venus's phases

Galileo Galilei was born on February 15, 1564, in Pisa, Italy.

supported the newer idea. This was a major breakthrough. Later astronomers discovered that the Sun is the center of the solar system, but it is not the center of the universe.

Astronomers besides Galileo also used telescopes to study Venus. Some watched as the planet passed between Earth and the Sun. This event is called a transit. Transits of Venus are rare. They occur in pairs. A second transit takes place 8 years after the first. The pairs are more than 100 years apart. The first recorded transit of Venus was observed in 1639. Two British astronomers observed the transit from different locations.

Astronomers did not observe transits of Venus by looking directly at the Sun.

Instead, they placed a telescope in a dark room and pointed it at the Sun. This projected a small image of the Sun against a surface behind the telescope. Venus appeared as a small black dot moving across the bright projection.

People learned a lot about Venus from these observations. Russian scientist Mikhail Lomonosov discovered Venus's atmosphere during its transit in 1761. He noticed an unusual ring of light along the planet's edge. He suggested that this was where sunlight was passing through a dense atmosphere. Later observations showed that he was correct.

The rarity of Venus's transits made observing one an exciting event. Irish astronomer Sir Robert Stawell Ball

This composite image combines several pictures of Venus's 2012 transit.

witnessed the planet's 1882 transit. He later wrote, "To have seen even a part of a transit of Venus is an event to remember for a lifetime, and we felt more delight than can be easily expressed."[3]

Venus's size and brightness make it easy to see. But thick clouds hide its surface. This makes it hard to study in detail

One of the first radar observations of Venus was conducted at the Goldstone Observatory in Fort Irwin, California.

with telescopes. In the 1900s, scientists used **radar** devices on Earth to look through the clouds. Radar provided the first detailed maps of Venus's surface. These maps included mountains and valleys. They also showed volcanic plains.

Venus has captured human curiosity since before recorded history. Ancient people saw it as a symbol. Observations from Earth revealed some of its secrets. But there was still much more to learn about the bright and mysterious planet. The invention of spaceflight in the mid-1900s led to even more discoveries.

Life on Venus?

Some astronomers wondered whether there was life on Venus. People could see that Venus was similar in size to Earth. They could also see that it had an atmosphere. They wondered whether there were living things hidden beneath the clouds. These speculations were mostly put to rest after radar revealed Venus's harsh surface environment.

CHAPTER THREE

EXPLORING VENUS WITH SPACECRAFT

The first spacecraft to successfully visit Venus was *Mariner 2*. It was also the first spacecraft to successfully visit a planet besides Earth. The National Aeronautics and Space Administration (NASA) launched it on August 27, 1962. *Mariner 2* reached Venus that December.

The spacecraft only had a limited amount of time to study the planet. This was because the spacecraft was not designed

Two engineers stand next to *Mariner 1*. This spacecraft, identical to *Mariner 2*, failed during its launch toward Venus.

This illustration depicts *Mariner 2* sailing through space.

to orbit Venus. Instead, it flew by the planet. It got within 21,564 miles (34,704 km) of its surface. Then it continued on a path that put it in orbit around the Sun.

Mariner 2 collected lots of data during its limited time near Venus. It measured the planet's temperature and studied its clouds. The results showed that Venus was hotter and under more pressure than expected. These observations changed how people thought about Venus. *Mariner 2* was considered a success. Jack N. James was a leader on the *Mariner 2* team. He said, "There will be other missions to Venus, but there will never be another first mission to Venus."[4]

The first NASA spacecraft to orbit Venus was called *Pioneer Venus 1*. It launched into space on May 20, 1978. *Pioneer Venus 1* carried several scientific instruments. One used radar to map Venus's surface. This revealed the highest point on Venus. It was the peak of a mountain in the Maxwell Montes range. *Pioneer Venus 1* also provided evidence that Venus has little or no **magnetic field**.

THE VENERA PROGRAM

The Soviet Union's Venera program led the way in studying Venus's surface. Between 1961 and 1983, this program sent several **probes** to Venus. The Venera probes were the first to land on another planet.

Pioneer Venus 2

Pioneer Venus 2 **launched shortly after** ***Pioneer Venus 1*****.** ***Pioneer Venus 2*** **flew directly into the planet. Before this, it released four smaller spacecraft that also fell into Venus's atmosphere. They collected data about the substances in Venus's air. Two of the spacecraft even survived the impact on Venus's surface. One sent back data for almost 68 minutes.**

They faced extreme heat and pressure. Some probes lasted only a few minutes. Others lasted more than an hour. They sent back the first pictures of Venus's rocky surface.

The first three Venera spacecraft launched in the early and mid-1960s. Scientists lost contact with all three while they were on their way to Venus. *Venera 1* and *2* are believed to have gone into orbit around the Sun. *Venera 3* is believed to have crashed into Venus. This made it the first spacecraft to reach another planet's surface.

Venera 4, *5*, and *6* were designed to collect data as they fell through Venus's atmosphere. These spacecraft revealed that the atmosphere is mostly carbon dioxide.

The first Venera spacecraft to send back data from Venus's surface was *Venera 7*. It was the first successful landing on another planet.

This life-size model of *Venera 1* is on display in a museum in Moscow, Russia.

Later Venera missions included both orbiters and landers. The *Venera 9* lander sent back the first images of Venus's scorched surface. Later landers took more detailed pictures. But not every mission was entirely successful. *Venera 11* collected data about Venus's atmosphere before it successfully landed on the surface. But the protective covering on its camera failed to open. This meant the lander did not send back images of Venus's surface.

The *Venera 13* lander lasted the longest on Venus's surface. It sent back data for 127 minutes. It was only designed to last 32 minutes. *Venera 14* was identical to *Venera 13*. It lasted on Venus's surface for 57 minutes. The final two Venera spacecraft were orbiters. They mapped parts of the planet's surface before running out of fuel.

OTHER MISSIONS TO VENUS

NASA launched the *Magellan* spacecraft on May 4, 1989. It went into orbit around Venus. *Magellan* used radar to map the planet's surface in greater detail than ever before.

The spacecraft created detailed images of Venus's landforms. *Magellan* eventually produced the first complete map of Venus. The mission ended in 1994 when NASA scientists sent it into Venus's atmosphere. *Magellan* collected data about Venus's air before burning up.

The European Space Agency (ESA) launched *Venus Express* into space in

A model of *Venera 12* is on display in Moscow, Russia.

November 2005. It was the ESA's first mission to Venus. The spacecraft went into orbit around the planet in April 2006. *Venus Express* studied Venus's air and weather. It also studied Venus's faint magnetic field. The mission continued until 2014 when scientists lost contact with the spacecraft.

***Magellan* begins its journey toward Venus.**

Japan's *Akatsuki* spacecraft launched into space in 2010. The plan was for it to go into orbit around Venus that same year. But a malfunction caused the spacecraft to miss its chance. Mission scientists did not give up.

The team behind *Akatsuki* successfully put it in orbit around the planet in 2015. The spacecraft studied Venus's upper atmosphere. It was also designed to find lightning in Venus's clouds. In 2020, it may have spotted a flash of lightning. Scientists are not sure whether the flash was lightning or something else.

FUTURE MISSIONS

Scientists still have many questions about Venus. NASA has plans for a future mission to the planet. It is called *DAVINCI*. This mission will involve two spacecraft. One will fly by the planet twice. It will study Venus's atmosphere. And it will make new maps of the planet's surface.

The other spacecraft will fall through Venus's clouds. This probe will study the chemicals in Venus's atmosphere. The probe will also take detailed pictures of the planet's surface as it falls. It will continue to collect data if it survives the landing. Lonnie Shekhtman writes for NASA. She says, "The *DAVINCI* probe will be able to do things 1980s-era spacecraft couldn't."[5]

NASA plans to launch *DAVINCI* around 2030. After that it will launch another spacecraft toward Venus. It is called *VERITAS*. This orbiter will update the map

NASA Administrator Bill Nelson discusses the upcoming *DAVINCI* and *VERITAS* missions.

of Venus's surface produced by *Magellan*. It will also determine what materials make up Venus's core. Scientists hope that *VERITAS* will reveal more about how terrestrial planets like Venus and Earth change over time.

VENUS'S UNIQUE FEATURES

Several aspects of Venus make it stand out among the solar system's planets. Venus is one of two planets that have no moons. Mercury is the other. Scientists are not sure why Venus has no moons. It may never have had a moon. Or a moon may have fallen into Venus and broken apart billions of years ago.

Although Venus has no moon, it does have a quasi-satellite. This is an object

No moon is visible as Venus transits the Sun.

that appears to orbit a planet but actually orbits the Sun. Venus's quasi-satellite is called Zoozve. As it circles the Sun, Zoozve never gets too far from Venus. It appears to make a complete orbit around the planet as well. Zoozve's orbit is not stable. Its path around the Sun will slowly change. Soon it will no longer appear to be orbiting Venus. Scientists predict that this will happen about the year 2500.

Naming Zoozve

Before it was officially named, Zoozve was temporarily called 2002 VE. This name referred to the year in which it was discovered. Artist Alex Foster included 2002 VE in a poster of the solar system he created. But he mistakenly wrote the name as "Zoozve." The name caught on and eventually became official.

INSIDE VENUS

As terrestrial planets, Venus and Earth share the same basic internal structure. They both have cores, mantles, and crusts. But Venus and Earth are different in many ways. Earth has tectonic plates. These are huge slabs of rock that make up the crust.

Earth's continents slowly move into new positions over millions of years due to the planet's tectonic plates.

These plates slowly slide across the top of the mantle. This causes Earth's surface to change over time.

Venus does not have tectonic plates. Instead, it has one solid plate surrounding the entire planet. But there is evidence that it may once have had tectonic plates.

Scientists study the gases in Venus's atmosphere for a glimpse into the planet's past. Planetary scientist Alexander Evans explains, "The atmosphere may actually be the best way to understand some of the very ancient history of planets that is often not preserved on the surface."[6] Part of a planet's atmosphere comes from gases beneath its surface. This happens as plates slowly move. This can release gas from deep inside the planet. The

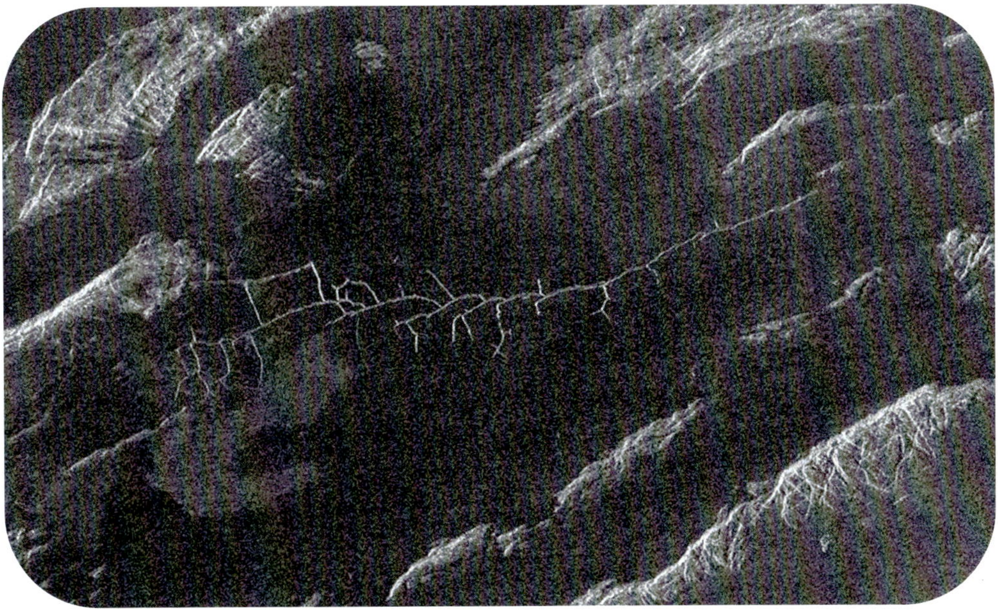

Although Venus does not have tectonic plates, its surface is far from inactive. This branching fracture on the planet's surface may have formed due to draining lava.

gases surrounding Venus suggest that the planet may once have had tectonic plates like Earth.

VENUS AND SOLAR WIND

Another unusual aspect of Venus is its lack of a strong magnetic field. Venus and Mars are the only planets in the solar system with this feature. Magnetic fields come from the movement of metallic material deep inside planets. Venus's lack of a magnetic field indicates that its inner structure differs in some ways from Earth's.

Magnetic fields shield planets from the solar wind. The Sun continuously emits solar wind. It has the power to tear away a

Solar wind is continuous, but sometimes the Sun emits increased bursts of solar wind.

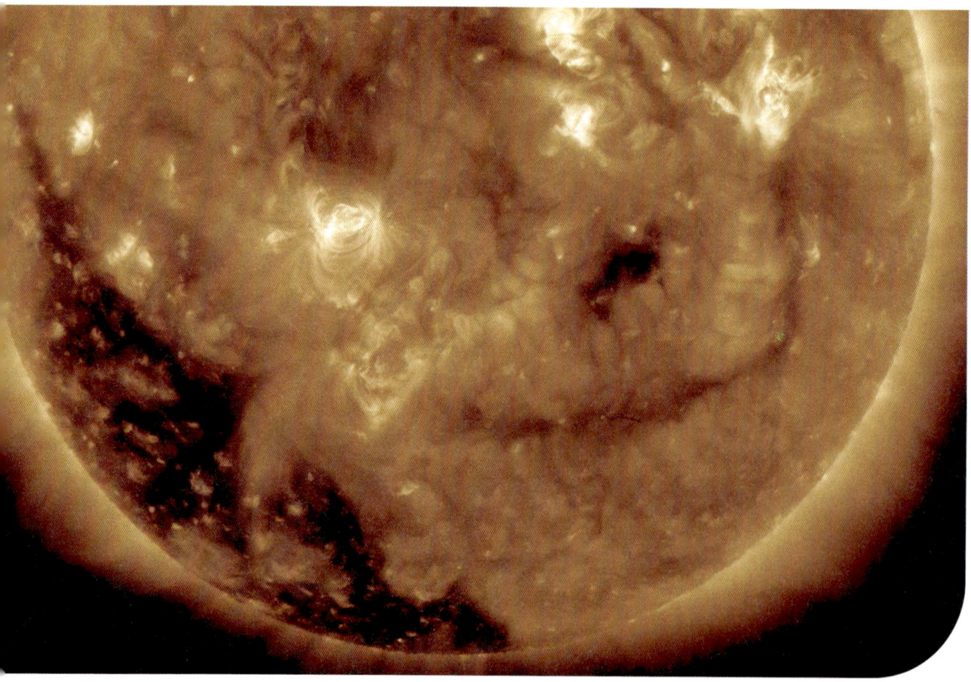

planet's atmosphere. But magnetic fields deflect solar wind. Earth's magnetic field helps the planet's atmosphere support life.

Without a magnetic field, Venus's atmosphere is exposed to solar wind. This interaction strips away lighter elements. This may have led to the loss of the planet's water over time.

SUPER-ROTATING WINDS

Venus's upper atmosphere is home to incredibly fast winds. These winds are known as super-rotating winds. They can circle the planet in just 4 Earth days. This is much faster than Venus's rotation. This creates a strange weather system. The atmosphere moves much faster than the planet itself.

One strange aspect of Venus's weather happens at the planet's poles. There, swirling winds form hurricane-like storms. The calm center around which a hurricane swirls is called the eye. Hurricanes on Earth typically have one eye. But sometimes a second eye can form.

Scientists understand why this happens on Earth. But they do not understand why the same thing happens on Venus. Storms at the planet's poles also sometimes display two eyes. This adds to the mystery of Venus's weather.

Scientists study these winds to learn more about Venus. They are curious to know how Venus's extreme environment formed. And they want to know why Venus and Earth are so different from each other today. Planetary scientist Sue Smrekar says, "We believe [Venus and Earth] started out with the same composition, the same water and carbon dioxide. And they've gone down two completely different paths. So why?"[7] Answering this question could teach people more about the future of their own planet.

This photo, taken with a special camera that captures ultraviolet light, reveals Venus's clouds.

GLOSSARY

astronomer

a scientist who studies space

axis of rotation

an imaginary line running through a planet's poles around which the planet rotates

greenhouse effect

a process in which the gases in a planet's atmosphere trap heat, making the surface hotter

magnetic field

the area around a magnetic material within which the force of magnetism works

phases

the different apparent shapes of a planet or moon as seen from Earth, caused by how sunlight hits it

probes

devices sent to space that gather information

radar

a tool that uses radio waves to detect a distant object's position

solar wind

a stream of charged particles released from the Sun

SOURCE NOTES

CHAPTER ONE: MEET VENUS

1. Quoted in Jules Bernstein, "Why Venus Rotates, Slowly, Despite Sun's Powerful Grip," *University of California, Riverside*, April 20, 2022. https://news.ucr.edu.

2. Quoted in William Steigerwald, "NASA's *DAVINCI* Explores Ten Mysteries of Venus," *NASA*, October 20, 2021. www.nasa.gov.

CHAPTER TWO: STUDYING VENUS FROM EARTH

3. Quoted in "Transits of Venus," *Royal Astronomical Society*, n.d. https://ras.ac.uk.

CHAPTER THREE: EXPLORING VENUS WITH SPACECRAFT

4. Quoted in John Uri, "60 Years Ago: *Mariner 2* First to Explore Venus," *NASA*, December 14, 2022. www.nasa.gov.

5. Lonnie Shekhtman, "NASA *DAVINCI* Mission's Many 'Firsts' to Unlock Venus' Hidden Secrets," *NASA*, December 16, 2024. https://science.nasa.gov.

CHAPTER FOUR: VENUS'S UNIQUE FEATURES

6. Quoted in Juan Siliezar, "Venus Had Earth-like Plate Tectonics Billions of Years Ago, Study Suggests," *Brown University*, October 26, 2023. www.brown.edu.

7. Quoted in Anthony Greicius, "The Return to Venus and What It Means for Earth," *NASA*, December 11, 2019. www.nasa.gov.

FOR FURTHER RESEARCH

BOOKS

Carolyn Bennett Fraiser, *Mercury*. BrightPoint Press, 2026.

Gail Radley, *The Space Encyclopedia*. Abdo Publishing, 2023.

Henrietta Toth, *Robotics in Space*. BrightPoint Press, 2023.

INTERNET SOURCES

Nancy Atkinson, "The Quest for Life on Venus," *The Planetary Society*, March 2021. www.planetary.org.

William Steigerwald, "NASA Study: Massive Volcanism May Have Altered Ancient Venus' Climate," *NASA*, November 17, 2022. www.nasa.gov.

"Venus," *National Air and Space Museum*, n.d. https://airandspace.si.edu.

WEBSITES

NASA: Eyes on the Solar System
https://eyes.nasa.gov/apps/solar-system

This NASA simulation allows visitors to explore the solar system on their devices. Clicking on planets such as Venus provides additional information.

NASA: Venus
https://science.nasa.gov/venus

The official NASA website on Venus summarizes what people have learned by studying the planet. The website also details past, present, and planned missions to Venus.

The Schools' Observatory
www.schoolsobservatory.org

The Schools' Observatory contains articles on topics in astronomy, including Venus. A section of the website provides students with information about career paths in astronomy.

INDEX

IMAGE CREDITS

Cover: © NASA
5: © NASA
7: © Detlev Van Ravenswaay/Science Source
8: © NASA
10: © NASA
13: © NASA
14: © NASA
16: © NASA
19: © NASA
20: © Olga Danylenko/Shutterstock Images
25: © Erkki Makkonen/Alamy
27: © Lennamats/Shutterstock Images
28: © Ilia Baksheev/Shutterstock Images
31: © NASA
32: © NASA
35: © NASA
36: © NASA
40: © Armael/Wikimedia Commons
43: © Don S. Montgomery/US Navy
44: © NASA
47: © Bill Ingalls/NASA
49: © Mark Sowa/NASA
51: © Reto Stöckli/NASA
53: © NASA
54: © NASA
57: © NASA

ABOUT THE AUTHOR

Jane Parks Gardner has written more than sixty books for students. She loves exploring big questions about space, nature, and science. Some of her favorite topics to write about are volcanoes, ocean creatures, and faraway planets. When she's not writing, Jane enjoys hiking, traveling, and searching for shooting stars.